Sloth Dreams Publishing LLC
Published by Sloth Dreams Publishing
Sloth Dreams Children's Books
Pennsylvania, USA

http://www.SlothDreams.com/kids

ISBN: 978-1-9824-5035-9

Flamingos

WRITTEN & ILLUSTRATED
BY KERIANNE JELINEK

Flamingos are beautiful pink birds. They are part of the family called Phoenicopteridae (fon-ee-kop-ter-ee-day).

FLAMINGOS ARE IN THE CLASS, CALLED AVES, ANOTHER NAME FOR BIRDS. PHOENICOPTERIFORMES (FONE-EE-KOP-TER-EE-FORM-ES) IS THE ORDER CLASSIFICATION FOR FLAMINGOS.

FLAMINGOS ARE WADING BIRDS THAT LIVE IN AND AROUND WATER. THEY LIVE NEAR LAKES, RIVERS AND OCEANS.

FLAMINGOS' HABITATS ARE IN WARM, WATERY REGIONS AROUND THE WORLD. THESE BIRDS THRIVE IN ALKALINE OR SALINE LAKES AND FAVOR SPECIAL ESTUARIES WITH HIGH LEVELS OF THEIR FOOD SOURCE.

FLAMINGOS LIVE MOSTLY IN AREAS WITH NO VEGETATION, AND LIVE IN AREAS WITH EXTENSIVE MUD FLATS AND PLACES WHERE THEY CAN FEED. THEY CAN BE FOUND IN FRESHWATER AND SALTWATER.

THE REGIONS WHERE FLAMINGOS LIVE ARE EITHER TROPICAL OR SUB-TROPICAL LATITUDES NEAR OR SURROUNDING AREAS OF THE EQUATOR.

GUYANA
SURINAME
GUYANA FRANCESA (FR.)
Paramaribo
Kourou
Isla del Diablo
Cayena
Afobaka
Amapa
Delta del Amazonas
45° O
Sentido de rotación de la tierra
30
60° O
Belém
Alcântara
São Luis
Parnaíba
Fortaleza
Aracati
Rocas
Fernando de Noronha
Cabo de São Roque
Natal
João Pessoa
Campina Grande
Recife
Maceió
Amazonas
Negro
Manaus
Madeira
Parintins
Santarém
Itaituba
Tapajóz
Xingu
Isla de Marajó
Imperatriz
Marabá
Caxias
Teresina
Crateús
Parnaiba
Tocantins
C A D E L S U R
Porto Velho
Pedro Afonso
Caruaru
Juazeiro
B R A S I L
Francisco
Feira de Santana
Aracajú
Alagoinhas
Nazaré
Salvador
Isla del Bananal
Gurupi
Barreiras
Alvorada
Bom Jesus da Lapa
Ilhéus
15°S
Itabuna
Mato Grosso
Guaporé
Aruana
Araguaia
Brasília
Cuiabá
Montes Claros
São
Goiânia
Anápolis
Caravelas
Diamantina
Governador Valadares
Coxim
Araguari
Corumbá
Uberaba
Belo Horizonte
Campo Grande
Paraná
Ribeirão Prêto
Juiz de Fora
Vitória
Campos
Bauru
Nova Iguacu
Petrópolis
Cabo Frio
São Paulo
Concepción
Paraguay
PARAGUAY
Ponta Grossa
Niterói
Rio de Janeiro
Santos
Santo André
REPRESA DE ITAIPU
Asunción
Formosa
Posadas
Lajes
Joinville
Florianopolis
Passo Fundo
Resistencia
Santiago del Estero
Salado
GRAN CHACO
Pilcomayo
Tarija
Santa Cruz
IVIA
Trindade

TYPICAL HABITATS WHERE FLAMINGOS LIVE INCLUDE: TIDAL FLATS, SANDY ISLANDS, MANGROVE SWAMPS AND INTERTIDAL ZONES.

Flamingos often alternate between legs while standing. Alternating back and forth helps them to regulate their body temperature. The "backward" bending area of the leg is actually the birds ankle. The knee of the leg is not visible, as it is tucked up under the flamingos plumage of feathers.

FLAMINGOS ARE OFTEN SEEN STANDING ON ONE LEG. THEY CAN DO THIS FOR UP TO FOUR HOURS AT A TIME. STANDING ON ONE LEG IS A NATURAL REFLEX THAT ALLOWS THEM TO MAINTAIN THEIR BALANCE.

FLAMINGOS HAVE LARGE, PINK, GRAYISH OR YELLOW FEET THAT ARE WEBBED. THEIR WEBBED FEET HELP THEM TO WALK IN MUD AND ALSO TO WADE IN DEEP WATER WHILE FEEDING. THEIR WEBBED FEET ALLOW THEM TO SWIM VERY EASILY THROUGH THE WATER.

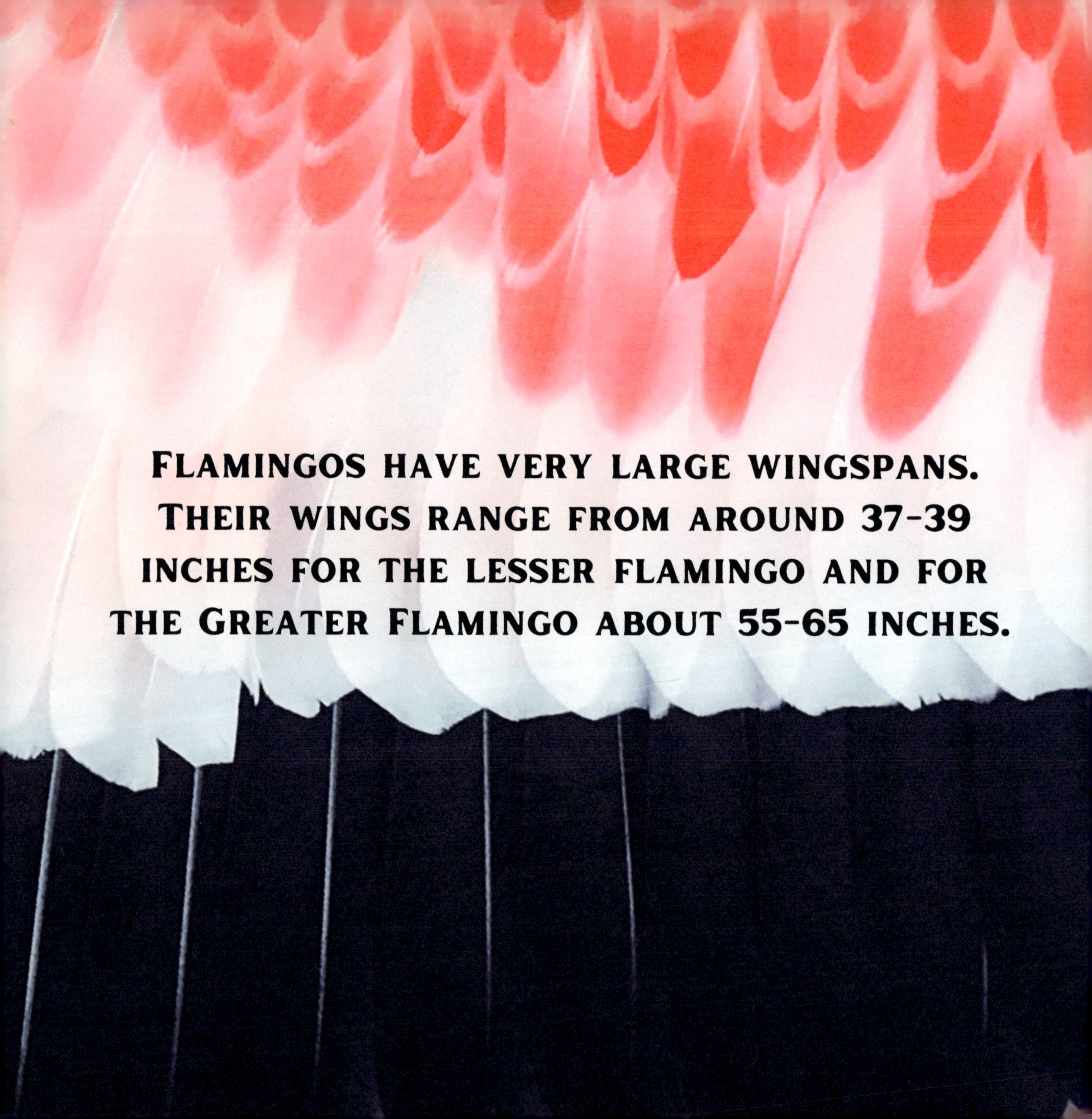

FLAMINGOS HAVE VERY LARGE WINGSPANS. THEIR WINGS RANGE FROM AROUND 37-39 INCHES FOR THE LESSER FLAMINGO AND FOR THE GREATER FLAMINGO ABOUT 55-65 INCHES.

FLAMINGOS DO NOT HAVE A SPECIFIC MATING SEASON. MATING COULD BE AT ANY TIME OF YEAR. TO ATTRACT MATES, BOTH FEMALES AND MALES JOIN IN ON AN ELABORATE DANCE. THEY DISPLAY THEIR LOUD CALLS, FLASHY FEATHERS AND WAVE THEIR HEADS AND NECK. THEIR DANCE IS FLASHY AND BRIGHT AND MUST BE A SIGHT TO BEHOLD. IT IS BELIEVED THE PLUMAGE COLOR IS A FACTOR IN HOW THE FLAMINGOS CHOOSE THEIR MATE.

A flamboyance is what a large group of flamingos is called. Flamingos are very social birds that live with many other flamingos in very large colonies.

FLAMINGOS CAN BE BRIGHT PINK, LIGHT PINK, PALE PINK, WHITE AND EVEN A BLUISH-GRAY SHADE.

Flamingos have a very specific diet. They eat brine shrimp and blue-green algae. The foods they eat give them their beautiful pink color. The pigment in these foods are called carotenoids. Carotenoids can produce orange, red and even yellow colors in plants and animals. Both algae and brine shrimp contain these pigments, thus turning the flamingos‘ feathers pink when metabolized by liver enzymes.

FLAMINGOS ARE UPSIDE DOWN FILTER FEEDERS. WHEN THEY EAT, THEY PLUNGE THEIR HEAD INTO THE WATER UPSIDE DOWN AND THEY SWEEP THEIR HEAD SIDE-TO-SIDE. THEY USE THEIR TONGUE TO SQUEEZE AND PUMP WATER OUT OF THEIR BILL. THEY HAVE COMB-LIKE RIDGES OR PLATES ON THE EDGE OF THEIR BEAK THAT CREATE SOMETHING LIKE A FILTER THAT WATER FLOWS IN AND OUT OF TRAPPING FOOD INSIDE THE BILL.

FLAMINGOS HAVE VERY UNIQUE AND LOUD HONKING VOCALIZATIONS. THESE NOISY BIRDS HAVE A VAST RANGE OF VOCALIZATIONS VARIATIONS AND NOISES THAT THEY USE TO RECOGNIZE THEIR CHICK, FOR MATING DISPLAYS, AND TO KEEP THEIR LARGE COLONIES TOGETHER. EACH SPECIES OF FLAMINGOS HAS THEIR OWN UNIQUE NASAL HONKS, GRUNTS AND GROWLS.

Flamingos make very large nest‘s out of mud, stones, branches and other plant materials. The nest will hold only one large egg.

FLAMINGOS HAVE ONE EGG EVERY SEASON. THE MOM AND DAD FLAMINGO ARE A MATED PAIR FOR THAT SEASON. THEY BOTH HELP TO SIT ON THE EGG, BUILD THE NEST AND INCUBATE THE EGG.

FLAMINGO EGGS TAKE ROUGHLY AROUND 27 TO 31 DAYS TO INCUBATE, AND TO HATCH. WHEN THE BABY CHICK HATCHES IT HAS SOFT DOWNY FEATHERS THAT ARE WHITE OR GRAYISH IN COLOR.

BABY FLAMINGOS ARE FED A THICK COTTAGE CHEESE LIKE SUBSTANCE CALLED, CROP MILK. THE FLAMINGO PARENTS PRODUCE THIS SUBSTANCE IN THE LINING OF THE UPPER DIGESTIVE TRACT. THE CROP MILK IS HIGH IN PROTEIN AND FAT FOR THE BABY CHICK.

A BABY FLAMINGO IS CALLED A FLAMINGLET OR CHICK. THEY BEGIN GETTING THEIR FEATHERS TO FLY AT AROUND 11 WEEKS OLD. AT TWO TO THREE MONTHS OLD, FLAMINGLETS TAKE TO THE SKY IN NUMEROUS ATTEMPTS TO LEARN HOW TO FLY.

FLAMINGLET'S STAY IN THEIR NEST FOR ABOUT 5-12 DAYS AFTER THEY'VE HATCHED. BABY FLAMINGOS MAY START ONLY TO LEAVE THE NEST WHEN THEY ARE STRONG ENOUGH TO WALK, USUALLY ABOUT 4-7 DAYS AFTER HATCHING. THEY ALWAYS RETURN TO THE NEST FOR FEEDING FROM THEIR PARENTS.

PARENTS OF THE BABY CHICK CAN RECOGNIZE THEIR BABY'S CALL FROM THEIR SPECIAL VOCALIZATION THAT IS ESTABLISHED BEFORE THE CHICK EVEN HATCHES. THIS CREATES A SPECIAL BOND BETWEEN THE BABY AND THE PARENTS.

After the baby birds can walk and venture outside of the nest, they join other flamingo babies in small groups. At about three weeks old the parent's start to herd all of the juvenile flamingos into large groups called creches. In the creche each baby bird starts to learn how to look for food for themselves.

FLAMINGOS LIVE FOR AN AVERAGE OF ABOUT 20-30 YEARS. MANY HAVE BEEN KNOWN TO LIVE UP TO 50 YEARS IN CAPTIVITY.

THERE ARE FOUR DIFFERENT KINDS OF FLAMINGOS SPECIES IN THE AMERICAS, AND TWO SPECIES THAT ARE NATIVE TO AFRO-EURASIAN REGIONS.

THE "AMERICAS" ARE SOUTH AMERICA, CENTRAL AMERICA, CARIBBEAN ISLANDS AND OF COURSE FLORIDA, IN NORTH AMERICA.

LOUISIANA
ustin
Baton Rouge
Tallahassee
New Orleans
FLORIDA
Orlando
onio
Houston
Tampa
St Petersburg
Miami
Nassau
Corpus Christi
Matamoros
Key West
THE BAHAM
Gulf of Mexico
Havana
CUBA
Camagüey
Tampico
Mérida
George Town
JAMAICA
Yucatan
Cayman
Islands
(UK)
Kingston
etaro
Puebla
Belmopan
BELIZE
Greate
axaca
GUATEMALA
HONDURAS
Tegucigalpa
Caribbe
Guatemala City
NICARAGUA
San Salvador
EL
SALVADOR
Managua
Barranqu
COSTA
RICA
San José
Panama
City
PANAMA

American Flamingos live in several countries throughout South America, Central America, the Caribbean Islands and Florida.

Specific areas in South America include: Chile, Peru, Argentina, Uruguay, Paraguay, Bolivia, Southern Brazil, Galapagos Islands, Ecuador and the Falkland Islands. Caribbean islands include: Cuba, Bahamas, Turks & Caicos & Trinidad and Tobago. Central America can include: the Yucatan Peninsula, Columbia &Venezuela.

FLORIDA
CENTRAL AMERICA
CARIBBEAN ISLANDS
SOUTH AMERICA
= AREAS FLAMINGOS LIVE

"AFRO-EURASIAN" REFERS TO COUNTRIES IN AFRICA, EUROPE AND ASIA.

Norwegian Sea
Arctic Circle
North Sea
Baltic Sea
Black Sea
Caspian Sea
Mediterranean Sea
Red Sea
Arabian Sea
Bay of Biscay
Gulf of Guinea
Equator
Libyan Desert
Nubian Desert
Sahara
Lappland
RUSS
SWEDEN
NORWAY
FINLAND
DENMARK
UNITED KINGDOM
IRELAND
WALES
ENGLAND
NETHERLANDS
BELGIUM
GERMANY
POLAND
LATVIA
ESTONIA
BELARUS
UKRAINE
MOLDOVA
SLOVAKIA
HUNGARY
AUSTRIA
SLOVENIA
ROMANIA
SERBIA
BULGARIA
MACEDONIA
ALBANIA
GREECE
FRANCE
SPAIN
PORTUGAL
TURKEY
GEORGIA
ARMENIA
AZERBAIJAN
KAZAKHSTAN
UZBEKISTAN
TURKMENISTAN
KYRGYZSTA
TAJIKISTAN
AFGHANISTAN
PAKISTAN
IRAN
IRAQ
SYRIA
LEBANON
ISRAEL
JORDAN
KUWAIT
BAHRAIN
QATAR
UNITED ARAB EMIRATES
SAUDI ARABIA
OMAN
YEMEN
CYPRUS
MALTA
TUNISIA
ALGERIA
LIBYA
EGYPT
WESTERN SAHARA
MAURITANIA
MALI
NIGER
CHAD
SUDAN
SOUTH SUDAN
ERITREA
DJIBOUTI
ETHIOPIA
SOMALIA
SENEGAL
GUINEA
SIERRA LEONE
LIBERIA
CÔTE D'IVOIRE
GHANA
TOGO
BENIN
BURKINA
NIGERIA
CAMEROON
CENTRAL AFRICAN REPUBLIC
EQUATORIAL GUINEA
SÃO TOMÉ & PRINCIPE
GABON
CONGO
DEM. REP. OF CONGO
UGANDA
KENYA
RWANDA
BURUNDI
TANZANIA
SEYCHELLES
MALDIVES

Countries throughout Asia include: India, Turkey, the Middle East, & Mediterranean.

Parts of Europe include: Southwest Spain, Portugal, Southern France and Eastern Siberia.

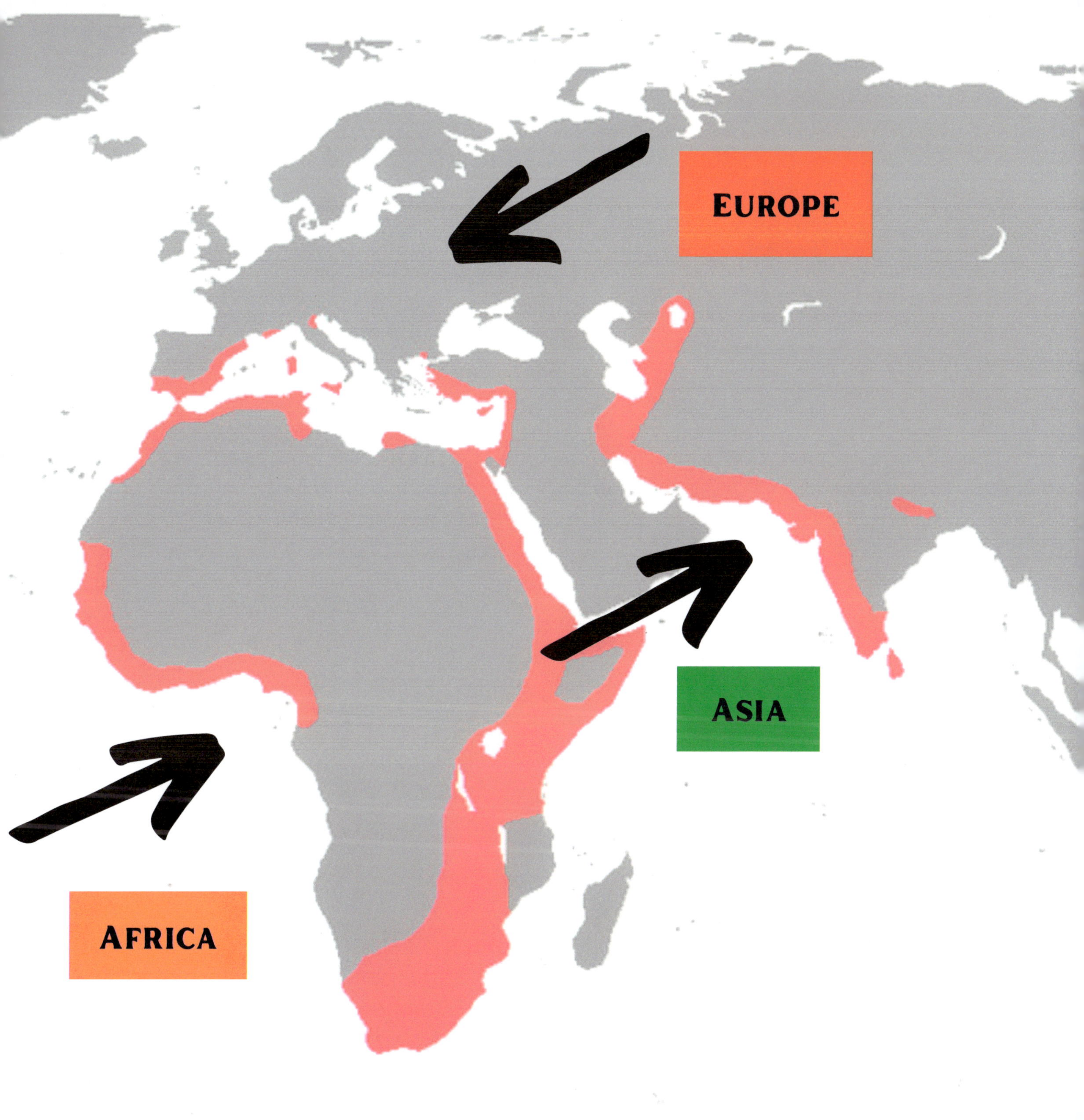
EUROPE
ASIA
AFRICA

THE AMERICAN OR CARIBBEAN FLAMINGO IS THE MOST POPULAR FLAMINGO. THEY CAN BE SEEN IN MANY ZOOS, AMUSEMENT PARKS AND ANIMAL PARKS AROUND THE WORLD.

THE FOUR SPECIES IN THE AMERICA'S ARE THE: CHILEAN FLAMINGO, JAMES'S FLAMINGO, ANDEAN FLAMINGO & CARIBBEAN (AMERICAN) FLAMINGO.

Caribbean Flamingo

The Caribbean Flamingo is mostly found in the Caribbean and parts of the Mexican Yucatan Peninsula. These birds are also sometimes found in Florida and even along the coast of southern Texas.

The Caribbean Flamingo is the largest of the flamingo species in the Americas. It can average around 60 inches tall and the males can be over 6 lbs in weight.

These flamingos are bright pink in color and have pinkish to gray colored legs. They have webbed feet. These flamingos have very distinct markings on their beak that are pink and black.

Chilean Flamingo

Chilean Flamingos live mainly in Chile, as its name suggests. They also live in parts of Argentina, Peru, Bolivia, Southeast Brazil, and Ecuador.

Chilean Flamingo have light gray legs with pink colored joints. Half of the beak of the flamingo is black. Their feathers are a very light pink color.

Chilean Flamingos have webbed feet and long stick-like legs.

James's Flamingo

James's Flamingos live in Bolivia, Chile, and in the highlands of Peru. The largest colony live in a shallow salty lake called, Laguna Colorado, located in Bolivia.

James's Flamingos are the smallest flamingo of South America. They are only 3 feet tall. Their bills are small and have a bright yellow color with a black tip.

These flamingos are pale pink, with very dark pink feathers on their necks and back. On their tails, they often have a few black feathers. They have a very distinct red patch around their eyes.

Andean Flamingo

Andean Flamingos are an endangered and vulnerable species of birds. They live in the Andean highlands, Argentina, Chile, Bolivia and Peru.

Andean Flamingos are the only flamingos that have yellowish legs. They are a relative to the James's Flamingo. Often, they share the same habitat. These amazing birds can travel up to 700 miles per day when migrating.

THE TWO SPECIES OF FLAMINGOS THAT LIVE IN AFRO-EURASIAN REGIONS ARE CALLED THE GREATER FLAMINGO AND THE LESSER FLAMINGO.

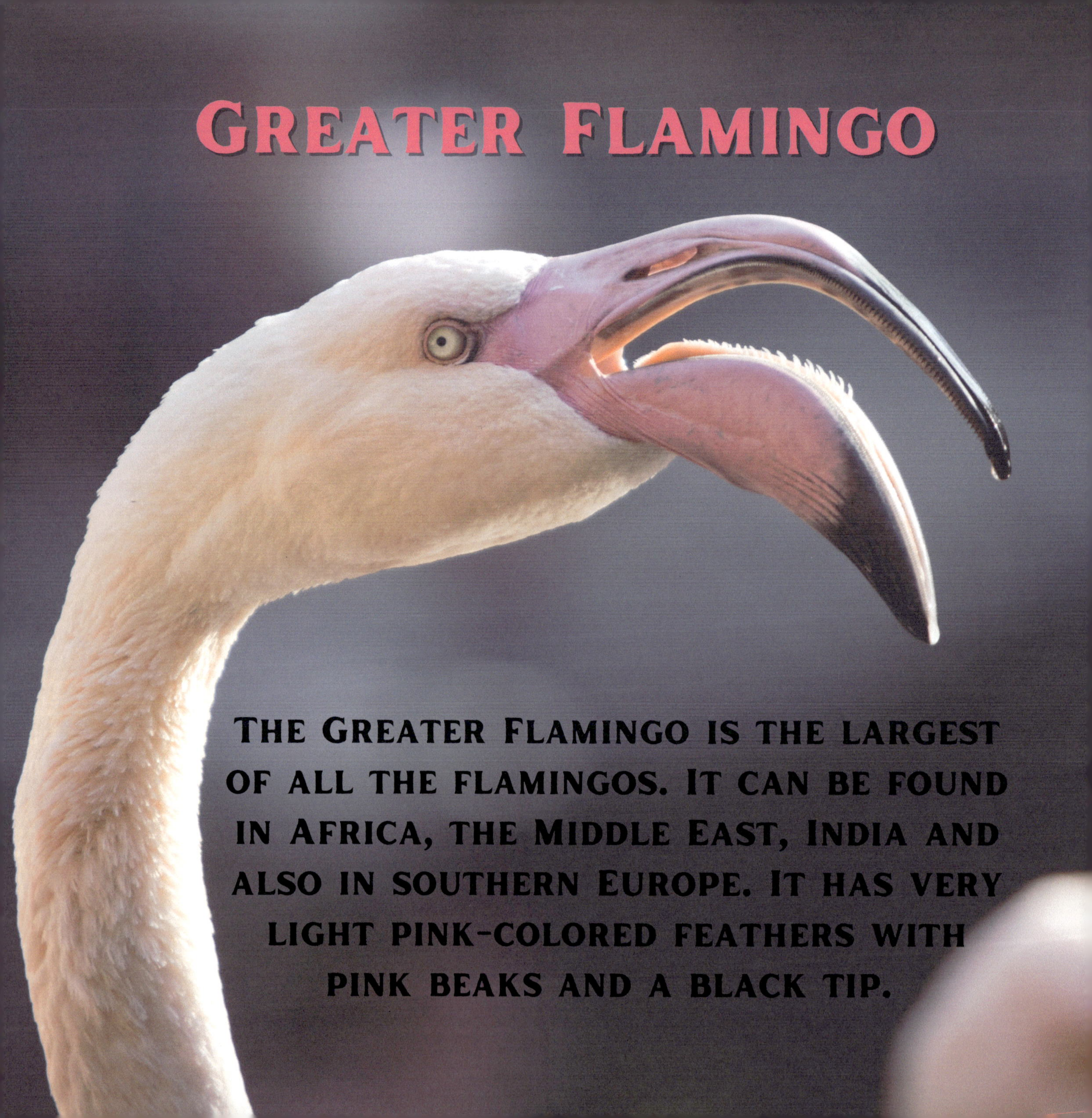

Greater Flamingo

The Greater Flamingo is the largest of all the flamingos. It can be found in Africa, the Middle East, India and also in southern Europe. It has very light pink-colored feathers with pink beaks and a black tip.

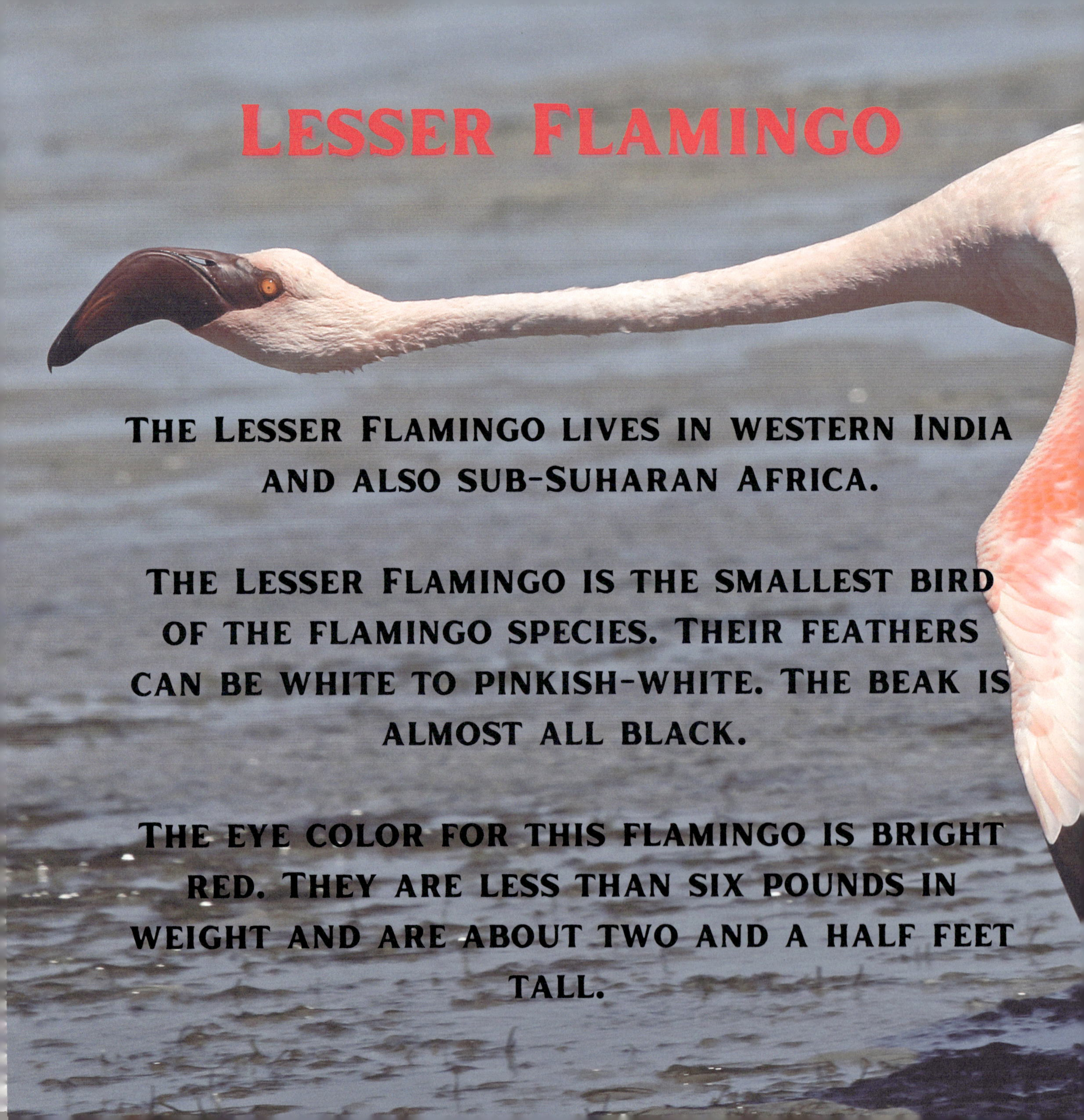

Lesser Flamingo

The Lesser Flamingo lives in western India and also sub-Suharan Africa.

The Lesser Flamingo is the smallest bird of the flamingo species. Their feathers can be white to pinkish-white. The beak is almost all black.

The eye color for this flamingo is bright red. They are less than six pounds in weight and are about two and a half feet tall.

AROUND THE WORLD FLAMINGOS HAVE DELIGHTED, INSPIRED AND FANCIED US WITH THEIR BEAUTY. FLAMINGOS HAVE BECOME POPULAR IN WORLDWIDE TRENDS, CLOTHING, OUTDOOR LANDSCAPING AND EVERYTHING IN BETWEEN.

FLAMINGOS ARE AMAZING ANIMALS THAT HAVE SO MUCH TO OFFER THIS BEAUTIFUL WORLD. MAY WE DO WHAT WE CAN TO ALWAYS PROTECT AND CONSERVE THESE MAGNIFICENT BIRDS.

Find Out More

To find out more about flamingo conservation and what we can do to make sure these animals survive, visit:

World Wildlife Fund
www.worldwildlife.org

Galapagos Conservation Trust
www.galapagosconservation.org

Wildlife Conservation Society
www.wcs.org

National Geographic Foundation
www.nationalgeographic.com

National Wildlife Federation
www.nwf.org

Visit Flamingos

Flamingos can be observed in their natural habits around the world and in zoos, animal parks and conservation centers. Here are some of the places you can explore flamingos around the world:

Paracas National Reserve, Peru
Renaissance Island, Aruba
Lake Bogoria National Park, Kenya
Lake Qarun, Egypt
Everglades National Park, Florida
Sardinia, Italy
Laguna Colorado, Bolivia
Great Rift Valley, Tanzania
Lake Natron, Tanzania
Lake Nakuru, Kenya
Key West Butterfly and Nature Conservancy, Key West, Florida
Inagua National Park, Bahamas
Flamingo Pond, Pumpkin Bluff Pond, Haulover Pond, Pigeon Pond, Lake Catherine, Turks and Caicos Islands
Flamingo Gardens, Davie, Florida
Busch Gardens, Tampa, Florida
Goto Lake, Washington Slagbaai National Park, Bonaire
Sarasota Jungle Gardens, Sarasota, Florida
Ardastra Gardens, Zoo & Conservation Centre, Nassau, Bahamas
Everglades Wonder Gardens, Bonita Springs, Florida
San Antonio Zoo, Texas
Smithsonian and National Zoo in Washington D.C.

REFERENCES

Koster Walton, C. (2019, June 18). Ever wanted to mingle with flamingos? Here are 10 ways to do it. http://www.10best.com/interests/outdoor-adventures/ever-wanted-to-mingle-with-flamingos-here-are-10-ways-to-do-it/amp/

Mayntz, M. (2020, February 2). Where to See Flamingos in the Wild - World Flamingo Viewing Hotspots. http://www.thespruce.com/where-to-see-flamingos-387032

Lavis, E. (2019, November 11). 6 Places to See Flamingos in the Wild. http://www.travelawaits.com/2484020/places-to-see-flamingos-in-the-wild/?amp

Wikipedia (2022) Lesser Flamingo. http://www.en.m.wikipedia.org/wiki/Lesser_Flamingo

Wikipedia (2022) Flamingo. http://www.en.m.wikipedia.org/wiki/Flamingo

Smithsonian's National Zoo & Conservation Biology Institute (2021, June 29). Why are Flamingos Pink? And Other Flamingo Facts. http://www.nationalzoo.si.edu/animals/news/why-are-flamingos-pink-and-other-flamingo-facts

SA Explorer (2013, September 23). South America's Flamboyance of Flamingos. http://www.saexpeditions.com/blog/post/south-americas-flamboyance-of-flamingos

SeaWorld Parks & Entertainment (2022). Habitat and Distribution. http://www.SeaWorld.org/animals/all-about/flamingos/habitat/

SeaWorld Parks & Entertainment (2022). Caribbean Flamingo. http://www.SeaWorld.org/animals/facts/birds/caribbean-flamingo/

Made in United States
Cleveland, OH
05 November 2026

47649169R00059